Wings on the Southwind

To my husband, John T. Daniel, who encouraged my expeditions to the Southern wetlands.

Wings on the Southwind

BIRDS AND CREATURES
OF THE SOUTHERN WETLANDS

THASE DANIEL

Text by FRANKLIN RUSSELL
Introduction by Roger Tory Peterson

Oxmoor House®

Library of Congress Catalog Number: 84-60286
ISBN: 0-8487-0546-7
Manufactured in the United States of America

First Edition

Editor in Chief: John Logue
Art Director: Bob Nance
Editor: Karen Phillips Irons
Production Manager: Jerry Higdon
Editorial Assistants: Cecilia Robinson, Lisa Gant, Pamela Hall
Associate Production Manager: Jim Thomas
Production Assistant: Jane Bonds

Pages ii and iii: Canada geese in flight over ragwort

Contents

Introduction   xi

The Eloquent Light   3

A Discourse of Birds   21

Rush to Feed   37

Blood in the Reeds   63

Grace in Breeding   85

Plants under the Wings   107

Fading Days   127

Farewell by Moonlight   143

OVERLEAF: *Water lilies under bald cypress trees*

Introduction

Do not be misled by her unusual first name—Thase. If you expect her to be a man, simply because wildlife photography seems to be a man-dominated sport, you are in for a surprise. This attractive woman from El Dorado, Arkansas, is one of the most skilled and versatile nature photographers in the country.

Thase Daniel did not set out at an early age to be a wildlife photographer. But, in a way, her photographic expertise is an extension of her musical skills. As a young woman, she received music degrees from Oachita Baptist College at Arkadelphia, and the Chicago Musical College. Later she taught piano at a girl's college in Shreveport, Louisiana. But she remained an outdoor woman at heart, having gone on hunting forays with her uncle since she was a girl.

It was inevitable that she would try her shooting eye with the camera. But her first attempt to photograph a bird—a steller's jay at a distance of 20 feet—was a humiliating disappointment. She recalls, "When the prints came back, I looked and looked for the jay, and finally made out this little black bug-looking thing way off in the distance." She soon realized that the camera, like the piano, is a mechanical device, a stubborn instrument with no mind of its own. One must make it behave, recognizing its limitations as well as its potential for creative artistry.

Louisiana bayou covered with duckweed

Thase, having mastered the intricacies of the keyboard, was not to be daunted by a mere camera. Noting her determination, her husband gave her a 35 millimeter camera, a 200 millimeter lens, and a trip to Arizona and New Mexico. She was off to a running start. She soon learned about the focal lengths of lenses, apertures, shutter speeds, film speeds, filters, and all of the other equations that go into basic photography. Added to this technical sophistication she brought the knowledge of woodlore instilled in her by her uncle. She acknowledges that pictures of friends and places are OK for the family album, but that photographing wildlife is far more of a challenge. Her love affair with the camera has brought her national recognition; her photographs have appeared in scores of books and in countless magazine articles. But in this book, a celebration of Thase Daniel's work, *every* photograph is her own.

Thase Daniel has been called an "expedition photographer." Although she spends a good share of the year not far from home, making sorties into the wetlands she knows so well, she likes to make at least one or two major trips each year to more distant places. "There is a great big world out there," she says, and although she loves the egrets, alligators, and white-tailed deer of the Southern swamps and marshes, she is equally ecstatic about flamingoes, penguins, polar bears, and other exotic wildlife. If she decides to explore a new and unfamiliar environment, she usually joins a nature-oriented tour such as one of those hosted by *Lindblad Travel*. Then, after an overview, she may select a certain hot spot and return to it on her own to concentrate in depth on its wild inhabitants.

I have had the pleasure of being with Thase on several of her photographic expeditions on the *Lindblad Explorer*. The Galapagos islands on the equator, 600 miles off the west coast of South America, is a wildlife photographer's dream. There I watched her expose literally hundreds of transparencies of frigate birds, boobies, albatrosses, pelicans, and other seabirds at scarcely more than arm's length. On one small island swarming with huge land iguanas and marine iguanas we found a short-eared owl so tame that we could photograph it at eyeball-to-eyeball range. In the U.S. or Canada this same species of owl usually does not allow a person to approach closer than 100 feet. The reason that the birds in this tropical Eden are so innocently fearless is that their populations evolved without the presence of that arch-predator, man, or any other threatening land-based mammal. Only in recent times has man, accompanied by his domesticated satellites, invaded this isolated archipelago.

Thase was not so fortunate on her return to the Galapagos. While composing a picture in the viewfinder of her Nikon, she was charged by a bull fur seal whose unexpected impact threw her on the rocks and broke her ankle in two places. While a fellow photographer distracted the furious animal, she crawled away, and she spent the rest of the cruise immobilized in her cabin on the *Explorer*.

I do not recall whether her camera was damaged in the mishap. I doubt that it was, because the first impulse of a wildlife photographer is "protect the equipment." On one occasion, when wading the shallow water of a saline lake in the Rift Valley of East Africa, she stepped into quicksand that threatened to engulf her. Being a real pro, she held the camera over her head while she struggled out. She knows that nothing will ruin a camera more quickly than saltwater or mud.

Like other professionals, she has learned that you win a few and lose a few. Not all expeditions are equally successful. I recall our cruise up the Amazon on the *Explorer*. Although the Amazon basin is reputedly rich in wildlife, you see little of it along the main river. Once beyond the flooded forests upstream from Belém, the countryside is monotonous. Almost all of the rain forest has been destroyed, replaced by marginal agriculture, and the traveler sees fewer birds than he would see at home. Although on our sorties by *Zodiac* on the tributaries we spotted a few parrots, macaws, and monkeys, they were not close enough for even our longest lenses. Thase, returning home with much of her film unexposed, decided not to repeat that trip.

East Africa, on the other hand, is tops in her estimation. The birds are abundant, colorful, and fairly approachable. And nowhere else in the world are the large herd animals and their predators so much in evidence. The semiopen terrain of the veldt and the bush is much more favorable for good film exposure than the rain forest with its hard contrasts of deep shadows and glaring patches of light.

To photograph birds and other wildlife, many photographers work from blinds made of strips of burlap or duck cloth sewn together and stretched over a jointed aluminum frame. The hidden photographer swelters inside with an eye glued to a lens of appropriate focal length which protrudes from a zippered opening. Why go to all the trouble of assembling and setting up an elaborate blind, Thase reasoned, when she could simply sit on a campstool with a double thickness of camouflaged mosquito netting thrown over her head so as to simulate the irregular contours of a bush? Good idea.

Thase sent me a few yards of the material that she picked up at a local Army and Navy surplus store in Arkansas and I tried it on the shore of Lake Nakuru in East Africa. My intended subjects were whistling ducks and flamingoes. Thase always feels safer with a pistol at her side when in strange places, but having no such protection, I felt uneasy when a small herd of wild buffalo came to drink. Fortunately, after I had exposed a roll of *Kodachrome*, they wandered down the lakeshore in the other direction.

Water hyacinths, lovely pests of the wetlands

Photographing wildlife can be an art, a science, a sport, or simply recreation—depending on the photographer. To me it has meant all of these things, but first and foremost it is a kind of therapy, a release from the tensions and pressures of my work at the desk and drawing board. I admit that I often use my transparencies as reference material or as a memory jog in my painting, but to be honest with myself, most of my photography is for pure fun. There is action and immediacy, quite in contrast to the sweat and blood I must endure during the long hours at the easel, while at the same time my visual acuity is sharpened. Both photography and painting force one to see a picture— but with a difference. Whereas a photograph is a record of a moment, a split second in time, a painting is a composite of the artist's experience, the past as well as the present.

Although Thase Daniel does not paint, she has an artist's perceptive eye; but I am sure that her obsession with photography owes its genesis more to her background as a hunter. Like so many sportsmen who have exchanged the gun for the telephoto lens, she likes the relative freedom of the camera. Although it requires more skill than firearms and may not reach out as far as a gun, there are not as many prohibitions and limitations; there are no closed seasons, no protected species, and no bag limits. The same animal can be "shot" again and again, yet live to give pleasure to others. It has been suggested that wildlife photography is really a remote survival from primitive times when nearly everyone had to hunt to stay alive. Millions still shoot for

Thase Daniel at work

sport, while others with a distaste for the taking of life (except for the table) subconsciously enjoy the thrill of the chase by bagging their game with a camera. A good picture is as tangible as a head of antlers on the wall or a stuffed grouse on the mantelpiece—and not as dust-catching.

Thase does not have to go far from her home in Arkansas to find her subjects; the wooded swamps and marshes of her home state are rich in wildlife. Or, she may pay her annual visit to her favorite heronry in Louisiana, where egrets and other long-legged waders congregate to breed.

Few areas in the world can match the wetlands along the Gulf Coast as a home for colonial waders and other water birds. Recently, with my friend Dan Guravich, a professional wildlife photographer who lives in Mississippi, I spent a fortnight in May on a photographic binge, visiting some of the hot spots between Biloxi, Mississippi, and Beaumont, Texas.

Mississippi, wedged between Alabama and Louisiana, has scarcely 60 miles of coastline, nearly half of which is man-made, dredged up from the Gulf of Mexico to form a fine sand beach intended for recreational purposes. The good citizens of Gulfport have reserved a full mile of this broad beach for the beleaguered least terns that attempted to nest there, and the struggling little colony of 70 or 80 pairs has now exploded to more than 5,000 pairs. Dan and I pulled to the edge of the four-lane highway that separates the beach from the motels, and I actually got frame-filling shots of the nesting terns simply by resting my 300 millimeter lens on the window of the car. That stretch of Mississippi coast is certainly the least tern capital of the world.

To visit the great colonies of royal terns and sandwich terns on the islets off the Mississippi Delta took a bit more planning. Inaccessible to the casual bird watcher, they are seldom visited except by the biologists and state wardens who monitor their welfare. Enlisting the help of these knowledgeable people who manage the wildlife resources of Louisiana, we traveled by boat through the marshy maze of the Delta to the Chandeleurs, a chain of sandy barrier islands far out in the Gulf. On our first landing our hosts proudly showed us a thriving colony of brown pelicans. The brown pelican, which is emblazoned on the state shield of Louisiana, is the state bird, but much to the concern of the citizens of the state, brown pelicans failed to raise young in Louisiana after 1961. They had become a casualty of DDT and other biocides in the food chain, chemical pollutants poured into the Gulf by the mighty Mississippi. Starting over again with young birds from Florida which were literally raised by hand, biologists sought to bring the species back, and now we are seeing the results of their efforts. It was estimated that in 1983 nearly 1,000 young were fledged in Louisiana. But unless their nesting islands are protected from beach erosion and high water, they may have serious setbacks in the future. Judicious dredging, using the spoil to build up the nesting islands by as little as two or three feet, could ensure their survival.

As wonderful as pelicans are, the big show in the Chandeleurs was the terns and skimmers. On one island we estimated 14,000 pairs of royal and sandwich terns, all crowded cheek-to-jowl in one massive colony. And that was not the largest aggregation. Scouting nearby Curlew Island by air, we spotted a colony which was later estimated at 40,000 pairs. We flew very high so as to create no disturbance and could only guess at the number of birds in each successive colony which showed up as a pearly gray mass against the pale sand.

Continuing our flight by small plane westward along the coast of Louisiana, we visited another prospering pelican colony before terminating our journey at one of the Audubon sanctuaries on the east coast of Texas. There we were treated to a glamorous display of egrets and spoonbills. Returning to my home in Connecticut I reflected on the advantages of living in the southern tier of states, so rich in aquatic wildlife.

Thase Daniel knows the rules and never disturbs colonial birds by entering their rookeries. She usually works along the perimeter, concealing herself with her camouflage netting during the cooler hours of the morning. She avoids the heat of the day, which could be disastrous to eggs or young. But her greatest challenge is to stalk her subjects while they are feeding or at rest.

The Southern wetlands are fragile: They will not take abuse. The birds, mammals, fish, and other wildlife that depend on the swamps, marshes, streams, lakes, bays, and beaches are a valuable resource, whether viewed from a purely economic standpoint—for example, commercial fishing—or recreation such as hunting, bird-watching, or photography. Wise management, already being carried out by the Fish and Wildlife Service of our federal government and the various state fish and game commissions, can be augmented by industry and even by individuals.

Wildlife photographers like Thase Daniel have done every bit as much as the nature-oriented writers in recent years to make the world aware of our fellow travelers on Planet Earth. In so doing, they have helped lay the groundwork for the environmental movement.

Keep up the good work, Thase, and watch out for those alligators and water moccasins!

Roger Tory Peterson

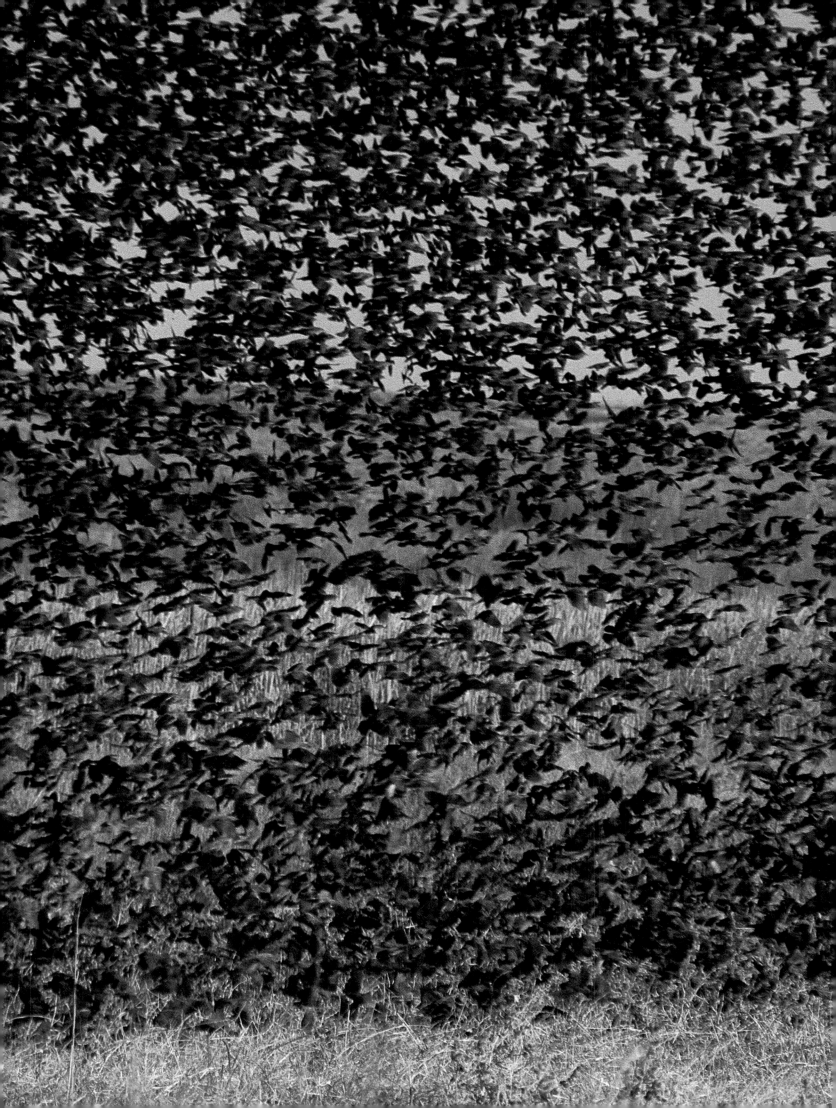

Wings on the Southwind

A curtain of blackbirds in Alabama

The Eloquent Light

In the early morning light, which supports a clumsy rail flying across a Louisiana bayou, a spider's web shimmers a necklace of jewels as each dewdrop is transformed into a prism. All light in these Southern wetlands has a touch of magic in it. But it is birds which transform the light into eloquence, a statement about life itself.

A snowy egret appears sculptured from white marble, until it raises its wings to form double parasols that look as though they have been whittled from light that was temporarily solid.

The great Southern wetlands, stretching from the tip of Florida to the southwestern shoreline of Texas, are like museums. At this early moment of the day, there is a pause that allows the exhibition of a unique American environment, often seen, but very little known outside of the South. From out of its stillness and quiet steals a sense of the strange and beautiful—a watchfulness of yellow-crowned night herons, an airy glide of white ibis, a weirdness of roseate spoonbills, a tininess of least bitterns, a grotesquerie of flamingoes, a thunder of geese and a gabble of ducks, a sibilance of warblers, and a great, high watchfulness of herons and eagles and turkey vultures.

From the earliest times, this sense of distinction has caught all travellers in its grip. In 1774, Robert Bartram wrote about abundance

Spiderweb in morning light

which is an echo of the past, but also a reminder that the causes of such abundance are still there, still creating a fertility of earth which cannot all be used up.

"The river . . ." he wrote, presumably of the Suwanee, " . . . from shore to shore, and perhaps near half a mile above and below me, appeared to be one solid bank of fish . . . pushing through this narrow pass of the Saint Juan's into the little lake . . . and . . . the alligators were in such incredible numbers, and so close together from shore to shore, that it would have been easy to have walked across on their heads, had the animals been harmless.

"Whilst this mighty army of fish were forcing the pass . . . thousands, I may say hundreds of thousands of them were caught and swallowed by the devouring alligators . . . the horrid noise of their closing jaws, their plunging amidst the broken ranks of fish, and rising with their prey some feet upright out of the water, the floods of water and blood gushing from their mouths . . . ."

The wetlands are museums in the sense that they are so well preserved despite the lumberman's axe (the Okefenokee Swamp has been cut from end to end, but still looks primeval), despite the periodic devastating fires (the Great Dismal Swamp once burned for six months), despite the hurricane and the tornado, the drought and thunderous summer rains that can knock down forests.

They are places where the lushness of American nature is still on display, as it was hundreds of years ago, when white men settled there. Indeed, much of the Southern wetlands is as untouched as if half a million years had never elapsed through the molecules of cypress and tupelo, cottonwood and buttonbush, prairie and hammock, Louisiana chenier and Carolina swampland.

The constancy of these wetlands is the unchanging parade of birds. It does not matter in which direction the eyes are set, sharply upward to the quick disappearance of a Mississippi kite in a battue area near the great river, or the sky-packed thrashing of thousands of snow geese struggling for altitude in Arkansas, it is birds which hold the mind and later recall the aquatic extravaganza that spawned them into vision.

The vistas of the wetlands are uniquely, almost aggressively, Southern. The agony-wrenched cypress, rising from a sunken boll as big as a small house, rising out of Lake Verret near the Mississippi, is a study in lonely grandeur—until a thousand warblers, thrushes, flycatchers fall from the April skies and rest in it during their migration from South America.

One quick glance into the mirrorlike waters of almost any Southern wetland reveals the reason for its abundance of life and for its preservation in a land swarming with machines that drain the waters,

Snowy egrets: eggs and nestlings

fell the trees, and strip the shrubs, to say nothing of millions of people who demand space, firewood, fruit, hunting, decorative plumes, and the right to drive, fly, sail anywhere.

The teeming waters of the South have endured the draining of a continent and that is the first miracle and fact of them. In Louisiana alone, there are nearly 12,000 square miles of bayous and bays, of pellucid lagoons and rippled lakes, of murmurous salt marshes and expanses of rolling rivers.

Any equation of nature that might be imagined is therefore available, superficially visible, then, looking down into such waters, as larvae of insects, tiny fish, frogs, snakes, toads, seeds, shrimp, crayfish, bass, bream, wriggling alligator youngsters, and a density of debris from vast rookeries of breeding water birds—eggshells, excrement, and the bones of those already eaten.

OVERLEAF: American avocets stop off in Louisiana marsh to feed and rest; then they fly to South America for the winter.

Lesser yellowlegs flies at great speed; they winter from central United States to southern South America.

But why a wonderland of birds?

Part of the answer lies in the words of an old Georgia swamp man, Joseph Hickox, who 20 years ago said to a friend from West Virginia, "Don't let your dog chase that bobcat."

He was saying, simply, that no domestic dog could come out of that Southern wetland alive. Yet the place, not a dozen miles from Waycross, Georgia, was hardly virgin wilderness. But it did teem with dog-loving alligators.

In Louisiana, the same idea was expressed by George Malcolm, a wildlife biologist, who told a Northerner friend wandering the salt marshes at Trevenick, "Leash the dog. There are wolves ahead of us."

The Southern wetlands have survived, in part at least, because they have been, and are, three things at once: inaccessible, fecund, and dangerous.

The long-legged birds that are ministers to what has been called "the religious experience of these wetlands" have endured triumphantly, for the greater part, because they have always been able to

Alligator takes the sun.

Tricolored young herons walk around on nests in South Carolina; they cannot fly yet.

Eastern screech-owl and his prey, a mouse

Great egret spreads her plumes in Texas.

White ibis nest in penthouses.

Roseate spoonbills perch near great egret nest.

Pair of great egrets (left) display, showing ruff neck feathers.

distance themselves from arrow, shotgun, and net behind thickets of vegetation filled with mosquitoes and poisonous snakes. Only when such tall and visible birds, such as whooping cranes, must journey out of the wetlands to breed do they become as vulnerable as other American birds.

The oldest heron rookery, located in a large cypress swamp on the South Santee River in South Carolina, has been there for at least 140 years. The Great Dismal Swamp is an unchanged voice from the very deep past. The Okefenokee of southern Georgia and northern Florida is missing only its otters, its ivory-billed woodpeckers, and its Carolina paroquets.

"Nothing but a skiff or dugout should be used to penetrate such places," wrote one Southern wetland traveller a century ago. "These can be propelled without a sound except for the sibilant dipping of a paddle. Silently, then, we enter a realm of tupelo gum, ash, and buttonwood, where the age-old calm is scarcely ruffled by a breeze in the treetops, the splash of a fish, or the song of a bird. A wide lagoon stretches before us, with aisles and leads forming passages through the ranks of tree trunks.

"There is an omnipresent sense of permanence coupled to a delicious diversity of form.

"Wide-winged ospreys float overhead. Reptilian anhingas perch on stumps, and an armored back, breaking the surface of the water, gives away the presence of an alligator. A brood of wood ducks threads through the brilliant green carpet of duckweed, and high over the trees, a swallow-tailed kite drifts in lazy circles."

The grunting notes of a glossy ibis, rising in sheeplike bleats, is a distinctive sound of eastern Texas. A grove of slender loblolly pines, rising like delicate jointless fingers, is a mark of Union Parish in Louisiana. The soft hiss of ibis wings in a twilit mangrove shore is atmospheric of the southwestern coast of Florida.

The spoonbills of the Gulf Shore are aloft, set against a murmurous gray sky suggesting squalls when the air is breathless and the birds turn toward a bayou. The early sun plays delicately on their extraordinary plumage, pink wings banded with deep carmine, tails a brilliant orange, all enhanced suddenly when the birds turn away from the eloquent light to avoid some towering live oaks.

The light swells, and dusky moss glows green, and a great white heron slowly raises magnificent fans of wings, which look nothing less than like Gabriel's wings, angelic flails to mount the heavens. The snaking neck, curved into a perfect S, ends in the downpointed, beady-eyed, dagger face, with feathered decoration mounted like a wisping handle behind the cranial dome.

Snowy egret nestlings and egg

A great egret lands on a tree limb near the nesting area.

Everything in the revealing light speaks of high specialization, almost an avian technology. The heron's long elegant legs, double-jointed, are themselves a design of an ancient adaptation in an unchanging place, thick-kneed, and also able to grasp a tupelo branch. The bird poses the question as to whether the sense of the aesthetic that led to human art came from the invention of art, or from watching the heron.

"A small lagoon opens to one side," continues the ancient chronicler, "and we turn into it to see the trees about its rim erupt into an explosion of sound and movement. Birds fill the air, wheeling, flapping, circling with discordant bedlam of squawks, shrieks, and now and then the gurgling notes of a snowy egret. Hundreds of nests, no more than platforms of sticks, are saddled in the lower growth. A few are low enough for us so that from our water-level position we can see the eggs—assurance of a new generation of Ardeidae to grace the green marshes and swamplands of our country."

16

The morning light lingers in palmettos, spills across the motionless plains of the Everglades and runs down into Mexico, across the Rio Grande. The scarlet ibis—like gorgeous pink fruit studding the rounded tops of mangroves—watch the sun's rise. A great blue heron stands, a perfect silhouette, on a Texas sandbar. A Louisiana salt marsh disgorges a flight of migrant warblers, recently arrived from Central America. A wild turkey cackles guardedly in the semitropical growth of a barrier island on the Carolinas shore. Moorhens and rails hesitate to begin their morning hunting at Jack's Creek, north of Charleston, and nearby freshwater ponds are choked with waiting ducks and geese. The clacking of clapper rails rattles across the salt marshes of the Eastern Shore.

Each bird is in its place, awaiting the day's work to begin. The Louisiana heron and snowy egret are fixed and content in their southeastern wetland regions. The great blue herons and green herons, once restricted to the Southern wetlands, are now spread between the Gulf and Canada. The black-crowned night heron has exploded from the wetlands and is found almost everywhere except in the deserts and higher mountains.

The wetlands are a genesis point, where the primevality is also the security of the place, which have allowed the common bitterns to also colonize much of the continent. Other Southern wetland birds are in the process of finding wider environments, such as the great egret and little blue heron which, once their wetland breeding season is done, in late summer, may filter up into New England and even to Ontario.

As a place of birth, the Southern wetlands teem. But they are also refuges for the needy, havens for the beset. They are a terminus point for those millions of waterfowl which filter South in fall and early winter. The hardiest birds stay close to the ice and snow, in Iowa, Oklahoma, northern Texas, and Missouri. The less tough birds—which include immature Canada geese, young swans, most of the ducks, and even some fish-eating mergansers—come to the wetlands to luxuriate in its abundances.

For migrants, the wetlands are a place of pause in both directions of their long journeys up and down the Americas. In spring, when cold fronts come out of Texas and cause severe squalls along the Gulf Coast, warblers, thrushes, and orioles may fall from the skies in clouds, and find temporary places among the egrets, herons, ibis, bitterns, and moorhens in the wetlands.

The last, and perhaps most unique, feature of the Southern wetlands is that they have been so thoroughly made a home by men and women who loved their seclusion and wildness so much they were prepared to lead lives totally unlike those of any other group of people in the Americas.

The Cajuns of Louisiana, political and religious refugees from Acadia, in Nova Scotia, not only settled the bayou lands of Louisiana, where nobody else would live, but have expanded and become an integral force in the life of the nonwetland parts of the state.

The Seminole Indians were like the egrets and herons and other long-legged wading birds in making the Everglades an ancestral home, until driven out, or exterminated, by the Spaniard, English, and American.

The swampers of Florida, Georgia, and other Southern states could not be separated from the natural history of the places where they lived—such as the Okefenokee—because they made themselves integral with it. They thought "Ameriky" was a terrible place where men were untrustworthy and the women painted themselves like savages.

Instead, they trusted the animals, particularly the birds, of swamp and marsh, prairie and lagoon, to guide them in their "primitive" lives. They listened intently to the cries of the barred owl which told them when various species of fish in the swamp were feeding and could be more easily caught. The hammering of woodpecker beaks enabled them to identify certain trees, which were navigation points, even when they could not actually see the tree.

They knew that each osprey's nest was about four miles from the next, and because each nest is different from another in shape, they always knew where they were in the vast interior of the Okefenokee. They matched their wits with the wild turkey, which they relished, and poled their ox-boats in pursuit of otters, which once thrived in the swamp.

From the air, the Okefenokee looks like trackless country, as does much of the Southern wetlands, a melding of pines into cypresses, of earth into sand, of peat into duckweed. Only the birds may get a sense of the vastness and grandeur of the wetlands, as they course back and forth across them, now excited and noisy in the eloquent light of this new day.

Royal tern and skimmers over Dauphin Island in Alabama

Vines, Louisiana iris, abandoned boat

A Discourse of Birds

In the wildest wetlands of Mississippi, Georgia, and Florida, there once lived a fabled swamp bird. It had a voice like a bugle, a shining white beak which looked like silver fired by the sun, and its hammering was so distinctive in swamp, hammock, and prairie wetland that Southern swamp dwellers navigated by the sound of it.

This was the ivory-billed woodpecker. Although it is now technically extinct, its spirit lingers. Its ghost is the presence of everything that is unique about the Southern wetland birds and the place where they live. The bugling cries are gone, but almost everything else that coexisted in that world remains.

How might this crow-sized woodpecker be made to exemplify tall wading birds, creeping rails and moorhens, showy waterfowl and soaring birds of prey?

The ivory-bill was most active in the hours immediately following the dawn and may be remembered as a trigger of Southern wetland life, the first bird to be heard or seen as the sun rises. The sound of this woodpecker, by all accounts, sent the blood surging through the veins of every Southern traveller in the nineteenth century, and before, who ventured from the red dirt of pine country into the quaking earths of cypresses.

A purple gallinule in search of insects . . .

. . . lifts lily pad with its bill and hunts insects.

"Delicious morning—green leaves, sweet smells, and an ambrosial breeze. Got a glimpse of the great bird while I was making a detour southward to pass between him and the swamp. Knew him by the sparkling white he showed and by the flare of carmine."

The traveller is Maurice Thompson, a nineteenth century naturalist, and the scene is purely Southern. The backdrop is memorable. A serried row of egrets stand, like ballet dancers, necks curved against the dense green of a hammock. In the air, a twisting funnel of cranes are rising in the soft morning wind. The sparkle of sun among whisk-flowers, nodding yellow blossoms on reflecting prairie wastes, also reveals a pair of lean, busy rails, spreading wide-clawed feet, each bigger than a dozen water lily leaves, literally seeming to walk on water, dabbing insect larvae from the surface.

The ivory-bill trumpets, and Thompson pauses in awe. But he is motionless for another reason. This is not the conservation-minded twentieth century. Thompson plans to kill the ivory-bill. He has an arrow unquivered, notched into his bow, and ready to make an addition to his collection of Southern wetland bird skins.

He fires the arrow. It raps harmlessly against bark. The great woodpecker, a little smaller than an eagle, flies away with its peculiarly undulating flight.

Between 1935 and 1938, the ivory-billed woodpecker struggled to survive in one limited area of Louisiana, the Singer Preserve, near Tallulah. It is just after sunrise, and a drizzling rain has begun on Christmas Day, 1934, and George H. Lowery, director of the Louisiana State University Museum of Science, is stalking the high-pitched, nasal YAMP, YAMP of ivory-bills.

Suddenly, he comes upon four of them, two pairs, smashing off the bark of a tree in search of flat-headed beetles, or "betsy-bugs." "They might have survived," Lowery recalled, "but by 1938, the last of the virgin hardwood bottomland swamp forest had gone because there was no sentiment to save it!"

Seventy years earlier, Thompson was more fortunate. "In the thickets and brakes which fringe the swamp," said Thompson, "I see many nests. Even the logcocks (pileateds) and ivory-bills choose the pine trees rather than the cypresses. Near the sluggish little streams, there are wild haw thickets. In these, I noted jays, cardinal grosbeaks, various thrushes, and many warblers. Of woodpeckers, my list contains ivory-billed, logcock, goldenwing, red-cockade, red-belly, down, hair, yellow-belly, and red-head. The belted kingfisher was abundant beside the streams and open ponds."

A pair of moorhens chase each other into a crowberry thicket, bright with the fruit of early summer, and tendrils of Spanish moss hang from a yellow bough.

It is early morning, January, 1914, and Will H. Thompson, Maurice's brother, is on one of his many Southern wetland expeditions. "It was midsummer," he wrote, referring to a visit he was making to the Okefenokee, "when vast flocks of geese, ducks, and all manner of wild fowl had fled to the North, yet the swamp was teeming with game and with wild things that were not game. The alligators had not been hunted and they possessed the swamp. The wild turkeys were extremely abundant, and the eagles, hawks, wildcats, and other things that feed upon the small birds and animals were hideously abundant . . ."

It is possible to stand, and just listen, and the wetlands will reveal themselves through voices alone. The ivory-bill was perhaps the most strident and distinctive voice, rendered as "poop-poo-peep-peep-poor-yehr, poo-pee-paa-pee."

A Bachman's warbler makes its buzzing noise, like a machine. A quick, frantic gobble, then silence, and a turkey is gone. The machine-gun-like rapidity of a hairy woodpecker drowns the thin whispering song of the myrtle warbler. Some sounds, heard again and again, never become associated with other birds. The king rail's asthmatic coughing and gasping is a bizarre counterpoint to the vomiting, squeaking voice of the purple moorhen.

The common loon gives its weird, insane cry from the dusk of the Mississippi in spring as it heads, with its mate, towards a Northern birch forest and its lakes where it will breed. The American bittern makes its voice into a mallet driving a stake in mud, then turns its bill upward—"the sun-gazer," say the locals—and pretends it has disappeared.

The old swamp people had their own names for the wetland bird residents, all of which had regional significance to each of the peoples. The Good-God woodpecker (black-and-white wings, scarlet crest) was heard along with the cham-chack and the white-shirt. The sandhill crane was a "whooper" to the Okie swampers, and the snow goose was an "eagle-head" to the Cajun people.

The environment is primeval, and so are both the sounds and the shapes. From somewhere far distant, a loud laugh comes rolling across waters, an almost derisive quality to it, the call of the king rail. The carpenter frog sounds right after the machinelike warble, its call a hammer rapidly striking metal. A pig frog grunts.

Even in winter, bird song remains brilliant and beautiful, but most particularly in late winter, when the birds are anticipating spring. By then, the wetlands have become almost drab. The cypresses seem dead. The grasses have collapsed and turned yellow, or gray. The air is leaden. This is the time for yellow-rumped warblers as they scatter liquid chips of sound across the wetlands.

Fulvous whistling duck and ducklings swim in a bayou.

All light in these Southern wetlands has a touch of magic in it.

Prothonotary warbler feeds insect to nestling in its nest in tin can.

Pair of wood ducks swim under fall leaves in Oklahoma.

The wetlands are ever changing. No day resembles yesterday.

29

Pig frog looks out from under a lily pad in Florida.

They are like yellow flowers darting forward, their rumps catching the low morning sun in fast flight, painting color in lines among the water plants, around the long black legs of the water birds. Like the red-winged blackbirds, they have a special affinity for the surface of the waters. They drop to snatch some unwary insect, landing on bonnets and lily pads, dashing from aquatic plant to low branch.

The wetlands are places of morning discourse. The density of vegetation may give concealment, but the distinctiveness of voice enables human observers to tell the season, almost the place, merely by standing still, eyes closed, and listening.

The heavy burring noise of great blue heron wings early in the morning might be heard with the faster, lighter wing beats of egrets,

A white ibis fishes in Wakulla Springs, Florida.

which it often accompanies to feeding grounds in the mornings. The great blues have always been sentinels, because of their wariness and their great height. This was a melancholy disadvantage during the height of unlimited duck hunting of years ago when the great blues were shot to ribbons by hunters frustrated in their efforts to surprise ducks in flocks of 30,000 and more.

The American egrets leave their roosts at dawn silently, but do not necessarily fly in flocks to their feeding grounds. They are individualists, and split as they fly, until each bird reaches a separate hunting territory, which might be in the same prairie, but not close to another egret.

The American black ducks, equally silent, watch the egrets carefully and have formed the habit of feeding in small flocks around stalking egrets on the understanding, apparently, that the tall and wary egret will warn them of danger long before any duck can see or sense it.

King rail wades in saline marsh, south Louisiana.

The ubiquitous gray catbird, with its yowling cries of alarm, bring other birds to the scene of any disturbance. Soon a dozen gray catbirds are in the area, then a Northern mockingbird, and then a red-shouldered hawk appearing suddenly in midair.

What might seem an empty and almost desolate wilderness is actually full of voices which may suggest hidden figures or convey information for which there is no sound.

The cries of the birds fade to silence. Six months later, in another season, the same place brings forth no gray catbird, no hawk, eagle, or water bird. The air is feverish, sticky, silent.

The wetlands of almost any morning are filled with the potential of surprise through voice. Sometimes, this is precisely the silence that is so dense it is a form of sound. But all such silences seem menacing, dangerous. When there is nothing to communicate, something must be wrong. In a moment, the speck of an eagle reports that some silences are deliberate.

In January mornings, the barred owls turn the wetlands into cacophony. These large, confident, and aggressive birds are well spread and visible, catching squirrels by daylight and crying out, deep, almost somber in resonance so that it becomes a kind of booming, a hollow drum announcing the sun. Then, a series of chuckles, screams, and shouts which conjure something more complex than an owl, more vivacious than a night bird.

On Honey Island, in the Okefenokee Swamp, the island may be cold, brooding, and silent to the human traveller. There is a grave dignity to its thick groves of pines. Then a choked cry comes from the trees, and a beautifully pileated woodpecker falls slowly, lands on its back on the pine needles, and dies.

Francis Harper, an ornithologist who listened carefully to the various discourses of Southern water birds, felt great nostalgia at the sight of the dead woodpecker. He cherished ancient memories in such

Barred owl perches among Spanish moss.

places. He listened for the Eastern wood-peewee singing its sad song, a bird that is spread across the wetlands as thoroughly as the gray catbird and the Northern mockingbird. He heard the Carolina wren and the chickadee in cypresses along the shoreline.

Kingbirds soon screamed at him as he stood there, near pines, and he heard a common yellowthroat, a battery of woodpeckers, and a patching of nuthatches.

The water turkey is silent and follows the traveller through swamps and lagoons. It has made all kinds of adjustments that have set it apart from other birds. It knows there is merit in being solitary, but that it is also safer to breed among the other communal birds of the wetlands, such as the ibis and the egrets. To join a common discourse might be dangerous.

But the sun rises into a swelling of voices. There is much to communicate. Frogs grunt and chirrup, and jump from the paths of early morning ox-boats. The Gulf horizons are patched with moving birds and lagoons in Alabama are incised with expanding circles of aquatic footprints left by just-departed herons.

A series of spine-tingling bugle notes smash the air of a Mississippi swamp. An island, which appears to be deserted at first glance, conceals a solitary sandhill crane, lurking in its shallows. Its voice is a melodious match for the roaring of alligators which will come later. It is a cry for which there is no real reason, except the sheer exuberance of there being communication and, perhaps, secret meanings to this morning discourse of Southern wetland birds.

The wetlands breathe, soft and deep, and the scream of an eagle is held in the white cup of a cloud.

Mallard drake bathes, tossing water with its bill.

Rush to Feed

At Anahuac, an idyllic refuge located between the Grand Seigneur Bayou, and the Pelletier Lagoon, in eastern Louisiana, fifteen thousand Canada geese raise their voices in a musical gabble. It is punctuated by a steady honking beat in the background. Many of the geese stand erect, beating their wings against the morning shadows of gnarled live oaks. Energy is building. Their feeding flight is about to begin.

The brilliance of the new day does not mean a willing start for many of the birds. It may be rather the reverse. The activity of the night has been so intense that the early morning is a series of pauses in which species after species readies itself for replenishment.

Or so it may seem. If the rails can be seen at all, it is a lean neck, narrow beak, small head projecting from marsh grasses, a bird that will walk to feed, and fly only if really desperate. If it believes itself unseen, particularly in the thick grasses of hay fields, it will allow itself to be run over by tractors and mowing machines, losing its feet or even its wings in its reluctance to fly.

The morning may be a time of hesitation springing from the fact that very few wetland birds are truly able to rest at night. There is too

OVERLEAF: Great blue heron flies over Okefenokee Swamp in Georgia.

much going on. Night is the meat-hunting time. Wolves, bobcats, raccoons, possums, bears are at work. The alligator is king of the waterways. The moon is cut by the flight of the big owls.

The Canada geese, though, like all flock birds, have a strategy of feeding. Their "plan" is to fly west, over Beaumont, Texas, and deep into the southcentral Texas rice fields. This food will last them, providing not too many birds join them from a November gathering near Brownsville. It will last if the Mexican Canadas do not come north when the December rains hit the east coast.

The feeding "aim" is to stretch this rice harvest over until mid-February when the birds will be almost ready to head back north for Manitoba, where they breed.

Then, in a streaming roar, the Canadas are aloft. The striving mass of birds rises above the live oaks and sends the ducklike grebes

Canada goose mirrored in the waters of the marsh

Canada geese reflections

Canada geese and goslings in ragwort

Least bittern fishes from water lily pads.

Common moorhen, once called the common gallinule

dropping to the bottom like so many sunken decoys. The grebes are as poor in flight as the rails, but their skulking escape lies in diving rather than being "lost" in vegetation. The geese go beyond the hammock, then the chenier at Pulaski, and so thence west into Texas.

All birds, whether the high-activity geese or the more deliberate wading and marsh birds, operate at high metabolic rates and cannot suffer starvation in the way, as an example, as can a bobcat or wolf. For the superhigh-metabolic birds, such as warblers, hummingbirds, and swallows, hours without food may bring death.

The mornings of the Southern wetlands, then, may be both a great deliberation before seeking food, or a near-desperate rush to hunt, to find food at any effort. When a cold front sweeps down the Florida peninsular, and flows over that natural masterpiece of the wetlands, the Everglades, it may upset the capability of many swamp birds to feed at all.

Tree swallows, which usually overwinter in the Everglades, south of Lake Okeechobee, are periodically caught in such cold-front advances, which usually overtake them at night, when they are sleeping in low Everglades vegetation.

The cold stiffens their wings and many cannot get into the air. The few hundred thousand that manage to fly immediately seek altitude to get above the cold wedge of air, when they turn east and head for the certain warmth of the Bahamas. Most of the other tree swallows will die.

The gadwalls of the salt marshes of South Carolina are as reluctant to move as the Canadas in Louisiana. They are used to feeding in the tidelands, between the so-called Sea Islands, and the saline estuarine marshes. But this is falcon, marsh hawk, and eagle country, and any large movement of birds will almost certainly become the object of attack. The gadwalls resolve this by leaving to feed in stages, a few hundred birds at a time, until the entire flock is dispersed, removed, and feeding distantly.

Suddenly, a thousand white ibis come planing down into a marshy shallow near Baton Rouge, onto broad tidal flats in the Mississippi Delta. Then, every frog, small fish, snail, even snakes are not safe from long, slender mandibles darting down at them like spears.

The shock of beginning the new day gives way to a kind of ebullient energy that sweeps all before it. Ibis are hunting for fiddler crabs on the Gulf Shore, with the silhouettes of oil derricks in the distance. They jab for crayfish in tidal estuarine shelters, competing with commercial fishermen for this crop, but also taking grasshoppers, cutworms, almost any insect that can be found. A four-foot wood stork, which might weigh ten pounds, will eat up to four pounds of snakes, frogs, fish, and insects every day.

Great egrets and glossy ibis fly from marsh after feeding.

The wetland birds spread beyond the marshes, out of the swamps and bayous, and are seen wandering fields, roadsides, and even airfields in this morning hunt for replenishment. The ubiquitous cattle egret, an involuntary migrant-settler from Africa in the 1950s, and which has become a bird of almost all environment, works very close to man and animal. It follows its African habit of staying close to domestic beasts, eating what is disturbed by the hosts' movements. But now, they are starting to follow commercial lawn mowers, snapping up the injured and the crippled from behind the flying blades.

The urgency of the great egrets to reach their feeding grounds is always according to the same ritual. Their hissing wings cut the slanted sunlight and carry them, thousands of crossed arrows, against the rising light. Oddly, they do not go to the same feeding grounds each morning. Instead, they seem to know exactly which grounds will provide the most food on any given morning.

The sun steals out of the live oaks and runs across a prairie, and the egrets come down like sport parachutists, wings flailing gracefully, to spread out in an immediate general massacre of fish, frogs, snakes, and insects. They can be seen, clumsy sprinters at best, running down

Cattle egret brings twig to nest.

The constancy of these wetlands is
the unchanging parade of birds.

Male Northern parula feeds nestling.

American avocet stops and feeds in a marsh in the spring on his way north.

mice and other rodents, through the shallows of their morning feeding territories.

The rush to feed is greatest in the wetlands in the months of April, May, and June, when many species are beginning to breed. Then, the two hours between six and eight are often uproarious with activity. Royal terns come inshore to the Sabine Refuge in Louisiana, and squabble in midair to reach the slugs that are crawling across the mud flats there, also bound on morning "rushes" to feed.

The skimmers, which live intensely and cannot always get enough food in the limited hours of early morning and evening when they can hunt, have gathered at the shores of Parry Island, where small fish are expected inshore along the beach line.

There, they congregate briefly with laughing gulls, as if gathering their own energies, like the geese, for the supreme effort of hunting. Suddenly, they take off together, spread out, separate, each bird taking a specific stretch of shoreline. Then, bottom mandibles lowered and just skimming a V through the wavelets, the oddly muffled wing beats of the skimmers sound in the fresh sun, like a child's footsteps running in deep sand.

The imagery of the Southern wetlands lays its presence in the imagination, even when every creature seems hidden or asleep. The morning movement to feed may be desperate for the swallows, play for the wood stork. Perhaps these tall wading birds were feeding in moonlight, or could not resist the lure of strong updrafts rising from the wetlands of southern Arkansas, western Georgia, central Mississippi, or the Carolinas.

The so-called wood ibis is the American wood stork, the only stork north of the Mexican border. Perhaps its soaring, high flights are as much a preparation for feeding as all this bustle down among the cypresses and cottonwoods, the prairies and cheniers of the glistening wetlands. The storks have risen thousands of feet and are soaring from an altitude where they can see the curve of the Gulf Shore, the grid-tracings of roads, the shrimp-fishing fleet returning from the night-hunt.

The storks fly for fun. It is exhilarating at this height. With a flip, some of them overturn and fly on their backs. At once, some of the storks start sideslipping, just as a pilot will do to reach an airfield or his carrier in a hurry. Down they come like dive-bombers, the rushing sound of wind, their half-closed six-foot wings becoming almost frightening on the ground until it is identified as the steeply-falling storks.

At the last minute, the birds stretch their wings; the roar of wind becomes thunder, and suddenly, they are all standing, motionless, ungainly now, in shallow water. Deliberately, one bird lowers its long spearlike beak, and a frog is impaled. Feeding has begun.

The spoonbills of Texas, or Florida, have taken off, pink wings banded with carmine, orange tails effulgent in the fresh light. A peregrine falcon zips out of a cloud and comes down at them at high speed. But both spoonbills and falcon seem to "understand" that this is just a game. The spoonbills turn away in leisurely change of course, and the peregrine modifies his dive into a flat, racing spurt that takes him out of range of the birds he might have killed.

The white ibis are uproariously busy in the mornings of their breeding in the Everglades, the nests studding tangles of gumbo-limbo and palmetto, up to a score of nests in every group. The piping of fresh-hatched chicks mingles with the gathering of thousands of adult

birds, readying themselves for feeding expeditions in marshy shallows, wet fields, or tidal flats along the western shores of the state.

But the morning expedition must be delayed for a moment. A black, teeming cloud of birds has appeared in the north, American crows and great-tailed grackles oddly mixed. The harsh nasal alarm cries of the ibis—runk-runk-runk—gargles outward. Hundreds more of the big birds take to the air which is soon filled with a confusion of honking crows, clattering great-tailed grackles, and swift, downward plunges of black-raimented birds to menace nests momentarily not defended by male or female ibis.

As abruptly as the attack began, it ends, and at once all the ibis are aloft, and the thunder of their wings, mixed with occasional farewell "runks," heralds their departure to feed and replenish.

Great egret fishing

Roseate spoonbill probes for food.

The Southern wetlands are very ancient places. That is evident from the uniform grace and habit of so many of its inhabitants. There is a superabundance of food, perhaps inexhaustible, but it is well spread, and success in numbers goes to those birds which have best devised tactics of discovery, so that they are always feeding where the food is thickest.

The ibis are the experts in this. But at the other end of the scale of "success" is the wetlands most primitive creature, the water turkey, or snakebird, the anhinga. It has not flocked with its comrades, which make no morning flights to feed, and which do not join any of the other wetland birds in any activity.

The anhinga is suggestively the Southern wetlands' link with the Age of Reptiles. It is called the snakebird, a strangely-endowed creature which can control its buoyancy, sink at will, so that only its neck protrudes from the water. Then it disappears to become an underwater arrow darting in search of snake, fish, or even small alligator.

Its rush to the morning hunt now occurs in a Georgia swamp, the Great Dismal, where, after a switching, darting passage through reeds and mangrove roots, it seizes an eight-inch-long alligator. In a moment, the reptile is ashore, pounded to death on a rock, and then swallowed. The anhinga's rush to hunt is over for the day.

The water birds move and bugle hollow cries, honk their way to secret places, ever moving in their sketching of the abundance that is all around but never certainly in one place. Against this, in the mid-morning of energy flights, some birds have long completed their feeding and are at rest.

The thin cries of Swainson's warblers fold across the mirrored waters of the Great Dismal. Quick snatches of melody from white-eyed vireos decorate recesses of the Okefenokee, and marsh wrens gurgle a kind of bubbling pleasure at having fed so well on insects made lethargic in the early morning chill.

A prowling falcon, invisible in Spanish moss camouflage, can look out upon Carolina wrens and tufted titmice, hear bursts of song from prothonotary warblers and their comrades, the Northern parulas and the hooded warblers.

A pinewoods sparrow is upon a hurrah bush, his neck craned upward to the thin, distant screech of a red-shouldered hawk, which is watching egrets on a prairie. The interaction of the morning movement to food is both opportunity for the meat-hunters and also a geography of how the location of food stocks has changed overnight. The shrimp have disappeared from a salt marsh channel in Louisiana. The crayfish are swarming in a Texas lagoon. Worms are surfacing in an Arkansas swamp.

53

Gulls fly after a shrimp boat in Gulf of Mexico.

Great egrets fish in a pond called Bird City on Avery Island, Louisiana.

Black-necked stilts pose like carvings.

A Florida sandhill crane walks near her nest.

The shrieking laughter of the barred owl, heard earlier in the morning as a late attachment to night, gives way to the male bird exploding out of an Alabama thicket carrying a wriggling gray squirrel in his talons. This is the owl which will decorate the nights of the wetlands, his screams and shouts and guttural chuckling laughter. There is eerie substance to the sound of death in the wetlands.

Red-winged blackbirds have gathered along the shore of a Mississippi levee, busy and gregarious, some using adjacent lily pads to dash

Purple gallinule stretches wings while feeding.

Common and Forster's terns fly from spit of land.

The white-winged birds sail
into the clouds in flights
that must be for joy.

Something excites feeding snow geese: some fly, some stand.

across the water as if truly walking on it. Distantly, something is laughing at them, and they all rise in a thick, noisy flock. A king rail will not reveal himself, except by this one rare cry, at any time during the day.

He will wait out the rising of the mosquitoes, the burnishing of a million ponds and lagoons and bayous and sloughs with a vertically-hot sun. He will watch the coming and going of water birds from rookery to feeding ground, and back again. But he will not be a part of it. Like the so-called skulking water birds—including rails, grebes, and loons—he is more a watcher, a prober, a wraith slipping to his destinations.

The morning's work is often interrupted, particularly where the sea is close. The interaction of Gulf and continental air is volatile. Many wading birds have not reached their feeding grounds when purple air boils up from the sea. The crash of thunder is heard from Aransas in Texas to the Everglades.

The morning thunderstorms of the Southern wetlands are most fierce in summer. But they can strike at any time, when temperature, humidity, and winds are right. These are true tropical downpours. Now, the Southern horizon is black for hundreds of miles. The rain advances like a wall into the wetlands and everything is blotted out.

The shock of the rain replaces the shock of awakening and now it is a time to endure an almost savage pouring down of water. A nuthatch clings motionless to a cypress trunk. A steady hail of debris, small branches, pieces of Spanish moss, small birds, come down to earth and water.

The hiss of water striking water in a murk that now envelopes everything is the sound of an army of snakes threatening an intruder. When lightning strikes, dim shapes of flailing egrets, or ibis, or spoonbills, or herons pass in jolting flight to shelters that may not be found before they are forced to land.

Now, although it is barely midmorning, it is as if the wetlands have returned to a time millions of years before the present. No artifact or art of man is visible anywhere now. Instead, the murk of rain invokes ghosts that are as remindful as the long-gone ivory-billed woodpecker.

They may include a magnificent frigate bird, brought down by the beat of the rain, after being driven north from the Caribbean. They might include cedar waxwings, floundering in marsh waters, unable to fly because they have so filled their throats with berries that they are nearly choked, close to "blacking out."

They may be refugees from very distant lands, such as a plague of dovekies, members of the auk family, which were driven into eastern wetlands by Arctic gales in the 1930s, thousands of refugees from Greenland.

A wood duck hen coaxes her nestlings to leave the nest box.

The rain pours down relentlessly and many birds, even including some of the true swamp dwellers, will drown. Most wetland creatures are precisely adjusted to their environments remaining fairly stable.

The tiny Bachman's warbler, first discovered in 1833 near Charleston, South Carolina, has such a slender hold on its wetlands habitat that heavy rains could wipe it from some areas. Its olive, black, and yellow form was not seen for 68 years after its first discovery, when it was rediscovered right where it had first been seen.

The murk of these great rains is a cloak for all mysteries of the Southern wetlands. Perhaps the ivory-bill still lurks in there somewhere. Perhaps a Bachman's is breeding in Alabama, Arkansas, or Missouri. Only a dozen nests of the bird have ever been discovered.

The rains hiss on, but they will be finished by midday. Replenishment may resume, or, by then, perhaps, it will be time for other kinds of hunting. Death will become more visible. Or it will not. The wetlands are ever changing. No day resembles yesterday.

Blood in the Reeds

The dark foliage of mangroves is the perfect backdrop for almost all wading birds. They stand, in their graceful silhouettes, before the gnarled limbs, the choked foliage, the tangled arms reaching up for a grip of air. Ibis, crimson as an old sun, become gaudily colored fruit when they land in the mangroves. The eagle, curled around a cloud, sends the ibis into instant hiding, the fruit withdrawn behind dark leaves. It is a disappearance of an invisible magician.

Perhaps this is the motif of the midday wetlands. The rush to hunt has peaked. Bellies and crops are filled, and there is time for introspective activity. Preening, transferring oil from tail glands to shaft feathers, is the work of hours in a day for a wetland wader. Motionless, apparently dozing inactivity is a silent induction of information from all around, a plotting of frog movements, a watching of other species' food plights, a scanning for the hurtling bird hunter.

Golden eagles and peregrine falcons once were relatively common in the Southern wetlands. Here were perfect ambush places for the falcon, poised like a coiled arrow on a branch camouflaged by Spanish moss, then snapped into action by the slow, clumsy rise of a heron seeking speed and altitude.

A dying Virginia rail hangs on barbed wire, unreachable across deep water.

The eagles, usually seen today only in singles or pairs, inspire a universal terror when they come upon rookeries of water birds. Once, they prowled the wetlands in hundreds, or thousands. The egrets, herons, cranes, spoonbills, and others are rarely taken by eagles today. But they must carry permanent collective memories of ancient massacres.

In such past experience is also contained both the ambiguity and the charm of the wetlands at midday. The dangers now are never so well concealed. But they also are never so real. Now, the victims stand very close to those who would kill them.

The situation is very similar to that of the high African plains, where grazing animals may munch grass all around killing animals—hyenas, lions, cheetah, and wild dogs—knowing when they are ready to hunt, and when they are not. The hunting birds of the wetlands have no difficulty killing to eat. Their problem is keeping their territories preserved so that each eagle, each falcon is not in constant abrasion with a competitor.

Today, experienced wetland watchers can expect to see many peregrine attacks as being more playful than lethal. A peregrine will place itself within easy attack range of a colony of freshly fledged spoonbills along the Texas keys. The youngsters all stand very erect and still, intently watching the raptor.

Then, with that characteristic jerk and drop of wings that signal the beginning of the peregrine's rush, the falcon takes off and comes straight for the young birds. Almost instantaneously, they duck their heads. The peregrine's passage is like a rapidly moving bowling ball, knocking down a series of peculiarly shaped pink pins. But no spoonbill dies. Within seconds, the falcon is gone. The mangroves seem empty. Then, the pink pins snap upward, as if mechanically propelled.

The responses to dangers are different for almost every species. The wood storks are nesting in young, ten-foot-high mangroves which have sprouted from the chaos of a hurricane that struck the Cuthbert Lake rookery region a dozen years before.

When the golden eagle slips between its clouds, and planes lower on broad wings, the storks demonstrate an almost monumental grace in the power of their wings, which force them upward into confusing, spiralling flight, from which the eagle turns, until the storks suddenly catch their updraft, rising from a bare glade, and go soaring, great gliders, more than half a mile high. They have escaped by flying toward the eagle.

The hunt steadily descends in drama from sudden death in the sky—there is a no more heart-wrenching sight than a wader hit by a

Portrait of an alligator

Osprey spreads its wings, ready to leave nest . . . then takes flight.

The wetlands are dangerous, and
every moment is precious.

Wood duck nestlings prepare to venture from their home.

falcon, and falling in a tangle of ruin—to pursuit in the mud. Spoonbills thrust up curtains of mud by their partially opened beaks slashing from side to side through the shallows in an Alabama marsh. Sensitive nerves in the inner lining of the beak tell the birds to snap their mandibles shut the moment a victim—small fish, insects, crustaceans—is sensed. The birds are so adapted to this method of hunting that they may hunt blindly, eyes closed, even when their prey is fully visible.

The midday wetlands slowly become places of tension. The white ibis colony in Florida seems inviolate, untouched by any kind of fear of danger, until suddenly some crows pass nearby. Near-panic follows. Harsh alarm cries—urnk-urnk-urnk—snort among the gumbo-limbos, and then soon die in the tangled vegetation when it is realized that crows no longer teem in the millions as they did a couple of hundred years ago, and made the Southern wetlands one of their favorite hunting grounds.

The midmorning heat is not steamy between November and March, as it may become in midsummer, when the most active hunting appears to be confined to early morning and late evening. Of course, it is only a delusion. The hunt is never called off.

A coyote, or a fox, or, in Louisiana, a wolf, eases itself slowly through haunted glades where the sunlight reaches only reluctantly. Dark and umbrous eyes of snakes and frogs watch, reptilian chill in their bodies. A cottonmouth hanging from a bough, drops with a plop as it sees the approaching, stalking mammal.

Perhaps this is a Texas key, or the shallow waters of Florida Bay. If it is Florida, then tangled roots hide many snakes, and the fox itself must watch carefully because poisonous snakes will strike at any disturbance.

A great blue heron which does not breed any further north than this mangrove country of southern Florida is somewhere ahead of the fox, unseen still, but betraying its presence with an occasional sharp, percussive slap of a dagger beak striking down, and the flip-flap sound of a struggling fat mullet, being speared, upended, and swallowed.

The thick foliage parts before the fox's muzzle, and a four-foot-high statue stands like a Japanese etching, watchful, but still unseeing of such a relatively small hunting animal. But danger is felt, rather than observed. The great white bird is supremely wary. With that odd slow-motion grace of all very tall water birds, the heron's body bows almost in two, the sweep of its wings expand, and takes bite on the air, the legs buckle and straighten, and the huge flying figure is aloft, ponderous, but in perfect conjunction with the smoky gray horizon, the reflecting waters, the dark slash of mangroves from which a pair of bright eyes watch, then disappear.

69

A nutria takes a rest on a duckweed-covered mound.

Nearly grown great egret cannot regain his nest in a cypress tree . . .

. . . *patient alligator consumes it in a gulp, feathers making a mustache on its long snout.*

71

Two young raccoons hide in an old stump.

Swamp rabbit out early to feed

A crayfish in his "chimney" made of mud

Red-eared turtle lays its eggs.

Beaver holds something in its paws as it swims.

74

Raccoon washes its food at the edge of a bayou.

A cottonmouth strikes twice, but the photographer was too quick.

Midday passes. The wetlands seem to sleep; illusion again. A minnow—of which there must be uncounted billions swarming throughout the South—darts this way and that, through mirrored waters that are watched by a thousand invisible eyes. It is not caught at once because there are so many minnows. Its life is dependent upon that accident of chance, when it conjoins with one of the tall, standing birds, with the mood of the osprey to drop, with the capability of the anhinga to come upon it suddenly in the reeds, with the desire of the raccoon to hook it out of the water with a swift scoop of the claws. The wetlands are dangerous, and every moment is precious.

A kingfisher drops unexpectedly and the minnow is gone. Another takes its place. It falls to a snake. Another takes its place. A bass eats it. Another takes its place. The green-backed heron, a little more than the size of a small crow, waits for the minnow, but not standing in the water as do most of the other water birds.

The bough upon which the green-backed heron rests is almost eighteen inches from the water. There seems to be no logical reason why it should be waiting there at all. Obviously, it cannot hunt anything in the water because it is too far from the surface. But this is the unexpected made predictable.

Almost all the pursued creatures of the wetlands have learned that one of the great indicators of danger is the twin black stems of waterbird legs. Perhaps the minnow cannot recognize this. But frogs and snakes can. The minnow comes closer to the bough. The beady eyes of the heron gleam in the post-meridiem sun, cruel, relentless, the striations of the body shifting radiances as the sun moves around it. Then, slowly, the heron's body goes into a crouch, the beak comes around so that it is pointed downward, toward the obliviously moving minnow.

The wetlands breathe, soft and deep, and the scream of an eagle is held in the white cup of a cloud.

In a movement too swift to follow with any ordinary eye, the green-backed heron drives its whole body downward, into the water and at the minnow. At the full extension of its body, it has performed the impossible. It has reached downward further than the length of its upright stance.

To do so, it has kept its grip on the bough with those powerful claws, so that the legs themselves are extended to give the body and the attack beak the full range of killing capability.

Then, with the minnow nipped in the tip of the beak, the entire apparatus of the hunt folds itself backward onto the bough, and becomes a green-backed heron again. The body becomes erect, but in the opposite way, vertically, and the fish is flipped so that its head is reversed (the great blue heron uses the same device for swallowing large fish), and is swallowed with convulsive gulps of both body and neck.

Young alligator climbs on its mama's head for protection.

Mother alligator hisses like a steam valve to protect the eggs in the nest (mound of grass) near water.

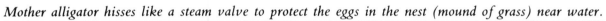

Little blue heron fishes in a pond in Arkansas.

The swamp rabbit must be as fleet as quicksilver as it quick-steps across lily pads to be one move ahead of the alligator. The raccoon, squeezing mud for crayfish, jerks its paw away, the paw bitten by its victim or, perhaps fatally, fanged by the cottonmouth. A sudden rush of waters, and the raccoon is dragged to deeper water by a big alligator.

The heron watches from above, the eagle is in its cloud, the pig frog sits on the leaf of a lily. The leaf is trembling, even though there is no current, no wind, no wave, in this secluded corner of the Okefenokee. But a large fish—perhaps one of the thousands of black drum that sometimes swarm up the Suwannee River—is nudging the leaf, attempting to force the frog to jump, its last act on earth.

The great blue heron hunts now. The statuesque bird advances, lifting each foot as though a drop of water might warn the prey of his advance. There is no ripple in the water. The foot is held, the bird as still as a rock carving. Half an hour passes. Beneath the mirrored reflection of cypress and live oak, loblolly and palmetto, a mullet waits,

A great egret in the Okefenokee Swamp, Georgia

perhaps seeing its enemy, perhaps not, equally dedicated to moving only when necessary, and remaining still for long periods.

Almost imperceptibly, the foot of the great blue is lowered, six inches ahead of the foot that is in the water. Underwater, it is the unfolding of joint and tendon, the placement of the foot so slow that not one mote of mud is disturbed. A crayfish, back into its mud shelter, is not even aware that the heron is there.

The heron sees the crayfish, but will not sacrifice its larger prey for the smaller. Great blue herons, coming upon concentrations of crayfish during the breeding season, have been counted eating up to 70 of the crustaceans, or between six and nine pounds of flesh, apiece. Despite such gorging, they could still take off and fly to a nearby hammock, where they rested, crops bulking, to digest.

The mullet waits, equally still. The two creatures appear to be double statues, one above, the other below, the water. The moment stands in time, and the murmur of insects, the dart of dragonflies, the appearance of a swallow-tailed kite over the peaks of distant cypresses, are continuum images that merely intensify the passionate moment of waiting.

The arrowed head of the great blue heron always appears to be fired from the long bow of its expressive body. The beak strikes diagonally downward, the speared point slipping through duckweed, plant stems, puffs of sediment, and clean into the mullet's midsection. It is always an explosion that seems furious, perhaps because of what invariably follows.

The fish is not dead but must be pulled to the surface with a jerk that often sees it slip clear. But if the spear is well rooted in backbone and gristle, up it comes, until it is almost five feet above the water. The neck of the great blue is extended vertically. Then the mullet is thrown into the air, seized again almost instantly, this time with the beak clamping the head, twisted, so that the fish faces downward. At once it is swallowed, in jerks.

Sometimes, the bird misjudges the size of the fish, or elects not to bash it to pulp before trying to swallow an extra-large victim. The fish sticks, halfway down. Now the great blue redoubles its efforts to swallow, great heaving gulps that seek to draw down the fish, regardless of size.

There comes a moment when it cannot go deeper. The side fins, which are spined, stick into the great blue's gullet. It is an impasse. Slowly, staggering from side to side, the great blue lurches in the shallows, perhaps towards land. It will take minutes, maybe an hour, before the bird chokes.

But frequently this kind of distress seems transmitted. Another kind of hunter is waiting and ready now to take both victims. The

Purple gallinule flies to a floating log.

water creams suddenly in a narrow furrow. The alligator has made its rush. Bird and mullet fall into an untidy pile, and the crunching of bone and feathers sounds like old dried sticks being broken up.

Great blue herons have been seen plummeting down into deep waters when fish are feeding at the surface, and floating, wings out-spread, striking with their dagger beaks. They have caused consternation, and reprisals, by fearlessly pirating fish hatchery pools. When the great blue is hunting, only the alligator—and it must be bigger than a great blue's throat—is safe.

The human hunter steps to front stage, both a villain and a character who has been an integral part of the Southern wetlands since the beginning of settlement. The crayfish, mud fish, catfish are the basis of a cuisine. The largemouth bass can still be caught, weighing fifteen pounds.

The water birds are never fully safe from catastrophe. The plume hunters have gone. But during the depression, fishermen began killing great blue herons in Florida Bay for food, reducing their numbers there from 1200 to 100. The birds recovered from that, but a 1960 hurricane killed sixty percent of their number in the Everglades National Park.

It is early afternoon, and there is blood in the reeds. The Southern wetlands await the next hunter with that silent certainty that is also the eloquence of survival.

The light swells, and dusky moss glows green, and a great blue heron slowly raises magnificent fans of wings.

Grace in Breeding

The mating dance of the snowy egrets creates the atmosphere for all breeding in the Southern wetlands. The feathers of the male are spread from his body, not as widely as a peacock's, but with an airy grace that makes them look less like feathers than some kind of man-made, wind-spread decoration. Even the chest feathers come away from the body, and protrude in a spiking design that is a part of the bizarre silhouette of a bird that is possessed in passion.

The female shows her beautiful plumage, too. The male, carried away, beats steadily upward a hundred feet, two hundred, three, then, almost hovering, like some outsized hummingbird, suddenly drops, tumbling end over end, and clearly about to crash at the female's feet in total disarray, if not in actual death.

But at the last moment, the bird rights itself a fraction before landing, and stands with his plumes raised perfectly in place, demonstrating his beauty, demanding of the female her attention and, presumably, her devotion.

In endless posing, against the hammocks, foots and knees of cypress swamps, the egrets appear to be natural works of art—gorgeous plumes in a gorgeous place. But it is less art than it is hard, practical

Cattle egrets display.

White-tailed buck races through a bayou . . .

. . . after a doe, whose four hooves fly above the water.

reality. These birds are playing out their secret of the wetlands; nothing here is by chance. It is all planned and organized, by the time of day, by the kinds of displays.

The snowy egrets are like some Japanese artist's conception of how wading birds might be idealized. The head crest is almost as spectacular as the nuptial plumes that spring from the birds' backs—not their tails. It was these feathers which so entranced the plume hunters of the nineteenth and early twentieth centuries.

Courtship in the Southern wetlands may occur in any month of the year and at any time of any day. Some wading birds, such as American avocets and lesser golden plovers, come into the wetlands in January and February, from the Caribbean and Central America, to breed. But regardless of species, all the wading brids—and some of the rails, moorhens, and waterfowl—have developed ritualized courtships which come near defiance of description.

The great egret, fighting with an adversary in the prairie under the sway of cypresses, the sun growling its long cry of heat into a nearby lagoon, is a picture by Matisse. Its wings are cast forward, like a woman's fan, the feathers spaced precisely, the sweep of the pinions producing the graceful hump of the straining back, supported under the backward-bent, thick-kneed legs, and claws. The spurs are six inches off the ground. Above, like a tower of grace, the long neck rises to the horizontally thrust spear of the beak, here used only in an emergency of action, not actually against the adversary, with whom there is no desire to damage, but merely intimidate.

Then just as abruptly as the attack begins, it stops, and the two birds stand erect again, facing each other, gurgling, and strutting, and squaring off their long wisping aigrettes, while the female stands nearby, demurely awaiting the outcome of the ritualistic fighting and display that is all for her.

The fighting of the breeding season rarely produces injuries. But it looks savage enough. The yellow beak of one bird suddenly slashes straight forward, directly for the eye of the other, like the rapier of a master fencer. The second bird jerks back deftly, and in the interregnum of the thrust, strikes back like lightning, only to see the head of the other bird dart to one side, and the beak pass its own eye with a quarter of an inch to spare.

When all fighting stops, a dozen males join in a solemn circular dance, like so many druidical figures, engaged in a mysterious process of passion that comes out of the mangroves, the cypresses, the prairies, the hammocks, the salt marshes, the very essence of the spirit within the great Southern wetlands.

After years of observing each of the water birds in detail, one Southerner, A. K. Herbert, commented that he had come to see them

Little blue heron on its perch

An American coot runs rapidly over the water to become airborne.

*It is a study in a matrimony that is
as old as the swamp itself.*

The sun highlights a white-tailed buck in its "velvet" horns.

Snowy egret fishes the waves in the Gulf of Mexico.

as people in disguise. There is so much about these birds that suggests a human direction of purpose that to anthropomorphize them is easy.

The whooping cranes of southeastern Texas, at five feet the tallest American birds, seem to become elongated, frenzied Russian dancers when they begin courtship. The almost equally big sandhill cranes walk around each other, heads raised, then lowered, then hopping high, raising and lowering the wings like blinds, and sometimes jumping over each other with guttural croaks.

The great blue heron does not grant a view of its courtship easily. However the big bird is approached, the atmosphere is always dramatic. The stealing silence, the faint mists gathering in mangroves, the tendrils of Spanish moss hanging, the giant bowls of cypresses and live oaks, the twitter of swamp sparrows, all are backdrop suggestions of what is to come.

In stalking the great blue heron, John James Audubon felt this atmosphere and wrote about it. "He has taken a silent step, and with

93

Female and male cattle egrets at nest

great care he advances: Slowly does he raise his head from his shoulders, and now, what a sudden start! His formidable bill has transfixed a perch, which he beats to death on the ground. See with what difficulty he gulps it down his capacious throat!"

Audubon's pictures of the great blue heron convey that peculiar combination of grace and fierce determination that so many of the Southern wetland birds display. The withdrawn neck is a prelude to a high courtship leap, the S-curve of the neck a readiness to strike at an intruder into the bird's territory, the long graceful body a strange contrast to the swordlike beak, which can spear a four-pound mullet.

The great blue leaps high again before his mate, and a delicate plume dangling from the back of the bird's neck stretches backwards, and flares, and seems to have no purpose except to make the courtship gesture artistically perfect.

Much of the courtship rituals must remain forever secrets, particularly among the skulking rails, coots, moorhens, and limpkins. They exchange gifts of small pieces of wood or fronds of vegetation, or the female limpkins receive the offerings of apple snails caught by the males.

The courtship period is thus a combination of both boldness and secrecy as each species seeks to establish that bond between male and female that will make breeding successful, and yet not so seriously expose themselves to enemies that they will be killed in attempting to re-create.

Yet it is also a time of much mortality. There is so much excitement in the wetlands at this time that all birds, secretive or not, become possessed, partially heedless. A Virginia rail gets carried away by her "feelings" and slams into a barbed wire fence where she is hung up, either to starve to death or be plucked away by an alligator.

The coot appears suddenly in the open prairie in Mississippi, itself filled with an emotion that the human watcher may never know, and scampers, flapping its skimpy wings. But instead of taking off, the bird describes a large circle, returns to the thicket from whence he emerged, and ducks back inside. Within, a female is hiding. Was his mock flight a gesture to her?

The purple gallinules stalk the edges of bayous, pausing from moment to moment to stretch their wings. At first it seems a series of inconsequential gestures. But soon, the care with which each bird lifts a lily pad with its beak, feet holding the leaf steadily, shows there is purpose in this, a kind of display. A female is always nearby, and watching.

The great egrets are at their nests, meanwhile, ruffing their neck feathers in similar purpose, even though, technically speaking, there is no longer any need to cement the bond that has been created between

94

Cattle egret with nestlings

Cattle egret on its perch

97

Cattle egrets fight over territory.

Dunlins feed in shallow water; they were once called red-backed sandpipers.

. . . people in disguise.

Wood stork preening in Florida

Northern bobwhites, cock and hen in Arkansas

them. The female settles on the nest. Slowly, the long aigrettes of the male are spread, diaphanous, a gesture so loving by human measurement that it is a shock to see the female unresponding, sitting, staring straight ahead, as though the display was not occurring, as though it were not specifically aimed at her.

The tricolored heron is seen stalking along alone, a tiny stick in her beak. She passes another tricolored heron sitting on her nest, and offers her the stick. It is struck from her beak with a sharp blow. Undaunted, she reaches down and picks it up again. The similarity of plumages in many wading birds is so great that the individual birds may not be able to tell whether the other bird is male or female until the voice is heard, a gesture is made, or a gift is offered.

When the water levels are right, snipe seem to become suddenly abundant in some parts of the wetland empire. Then, even though their occupancy of the South is only by winter, they can be seen as engrossed in courtship, and readying themselves for breeding as any of the residents.

When massed in hundreds, or even thousands, they are suddenly no longer secretive birds but bold in their displays. A swamper, Gad Roddenberry, saw several thousand snipe in a small prairie south of the canal dug in the Okefenokee Swamp, at Coffee Bay, in January 1934. They were in a state of high excitement, obviously in readiness for leaving for the North, destinations as far away as Newfoundland and Labrador.

Roddenberry, like so many wetland wanderers, was familiar with the early nuptial dances of the woodcock, which also begin in the late afternoon, also in the middle to the end of December. They are commonly seen in great numbers at Floyd's Island in the Okefenokee, rising into the sloping late afternoon sun, uttering their repetitive peent-peent cries, then flying in semicircles south over the hammock at the southern end of the island, and then flying north, twittering, toward the old logging railroad.

Nuptial dances, breeding, eggs, nestlings—the procedures begun by courtship may be seen in almost any month of the year in the Southern wetlands. It is at Aransas, Texas, though, that the best finale is made to the drama of the courtship time. The whooping cranes begin a late afternoon bugling of their excitement. The huge birds leap high on their territorial mounds.

It is March, and time to begin the flights north. This is the migration that brought them to the brink of extinction. If they could have evolved a breeding cycle that kept them in the Southern wetlands, they would be existent in the thousands today. But these bold birds, aggressive, demanding, and insistent on large areas of territory, simply did not have enough space in the South to hold them.

100

Killdeer stands over nest with four eggs.

Killdeer wades shallow water, repeating its name: "killdeer, killdeer."

It is midafternoon, and the sun is clouded in insects, hung about with humid vapors rising from steaming summer waters, a golden disk falling into the Gulf and leaving a trail of red in a thousand marsh wetlands.

The whooping cranes at Aransas are coming to the end of their courtship. This is the most solemn and significant of them all because these giant waders do not breed in the Southern wetlands. That is the reason why their numbers are so low, why they have come so close to extinction. Few of them are killed in the South. They die reaching for their Manitoba breeding grounds, and returning.

Thus, it is essential that the bond between the big birds created at Aransas be strong enough to last through the dangerous migration, ensure the capture of territory in the Canadian North, last throughout the summer breeding, and then bring both the birds back to the territory from whence they left. If the bond breaks, the birds either will not breed, or will die, or both.

Abruptly, the spring afternoons of the Aransas echo to bugling cries. One by one the huge birds rise and circle, each calling to a mate who might be a mile or more distant, rising from her own territory. It is a study in a matrimony that is as old as the swamp itself.

The sun is sinking, and the whoopers are disappearing into the north, into Oklahoma river lands, Nebraska ponds and rivers, North Dakota Missouri waters, dams, and levees. In days, they will be lost in the Canadian North, and their success in courtship, migration, and mating will not be known until they return to Texas in the fall.

*Eggs haunt the seasons,
millions, billions of them.*

Nestlings beg great egret for food.

Plants under the Wings

The plants of the wetlands reach out, support moorhen feet, give the slender rail a jumping-off place, enfold a decayed boat floating in a bayou, and provide a subtle, speechless background for tall birds stalking the front stage of the Southern theater.

It is plants, in all their Southern variety, that make the late afternoons of the wetlands such a splendid cooperative act between vegetation and water bird. The birds are always in the background, discreetly separating themselves from the prowling watcher. But the plants stand fast, can be taken in the fingers, smelled, seen in microscopic detail.

It is another kind of parade, hardly less distinguished than that of the birds, from duckweed to possumhaw, umbrella magnolias to cardinal flowers, that is the great instrument of change in the wetlands. The birds perform their routines, and are more or less predictable, but the plants are both ever changing and without warning of what they will do next.

James Walker, the manager of a commercial park refuge in the Okefenokee Swamp, used to take a canoe into the swamp early every morning. He said, "You never knew what you might expect each day.

Hooded warblers stand at nest in a viburnum shrub.

The formations of the swamp change. Entire lakes fill in suddenly as the trembling earth moves in. Or new lakes are formed as the floating vegetation is moved away by wind or currents."

This means an equally dynamic response of birds. Rookeries which have been in place for centuries might move almost overnight. Egrets, once common in an Alabama marsh, may disappear and cranes take their place.

"Large amounts of vegetation that have sunk sometimes rise abruptly to the surface," Walker said, "and vice versa, so that in exploring the swamp, you may become lost in a few minutes."

When Walker first came to the Okefenokee in 1955, he had never seen a sandhill crane on the northern side of the swamp. They were restricted to the Suwannee Canal Route refuge area. But, almost overnight, they started appearing in the northern swamp.

"I understood it was because of a plant, the paint-root, that the cranes love, which for some reason spread into the northern swamp in the late 1950s," Walker said.

The plants of the Southern wetlands are silent monitors and modulators of most bird life, indeed, all life. They provide the refuge, the food, the camouflage, the shelter, the nesting materials, and that sense of eternal serenity of which the birds, in their poses, their urgent flights to feed, their nuptial dances, are extensions and additions.

A dense swarm of cypresses, the product of a gale that caused a heavy seeding, rises from a bend in a marsh lagoon. Once, migrant water thrushes paused here to feed on molluscs in the muddy shores. Now, only creeping moorhens and rails can thread through such thick growth, beautifully pallid green, set among older, dark parent trees beyond.

The birds know the plants in ways that science cannot yet research, only suspect. The Indians revealed many of the plants' secrets, which might also be seen as knowledge shared by birds, but in different ways.

The never-wet, or golden club plant, thrusts up out of prairie waters. A thin coating of wax covers the tops of its floating leaves which, when the light is right, gives off a golden glow into the water. The Indians discovered there were tough, silklike fibers inside the leaf, and made their clothes from the never-wet.

Old swamp dwellers split the bonnet plant—the leaves of which they used as hats in summer—at the base of its leaves to get to a worm that lived in there. The two-inch worm, light green, made a magnetic bait for fish, swimming frantically only at the surface.

The birds watch the plants for both the insects that might be hidden among them and also for the abundant fruit that they supply, sometimes year-round. The tree huckleberry, or tree blueberry, carries

108

Cypress knees and white water lilies

Dragonfly lands on a water lily.

Solitary mauve flowers of the
pipewort take in the
colors of a sun falling into
the Mexican Gulf.

Lotus flower, bud, and seedpod in Arkansas swamp in July

its fruit deep into the winter if resident and migrating birds do not need the food. In some years, other plants, like the gallberry and the myrtle holly, also hold their fruit.

The wild turkeys come planing down over prairies brilliant with water hyacinths to scratch up mast laid down around live oaks in Bass River Refuge in south Alabama, while bears are climbing high to get at one of their favorite plant foods, the black gum berry, which endures into the winter throughout all the Eastern wetlands.

Here, plant and animal form a fantasy image, threaded among smoking shafts of sun that light everything but reveal nothing. The Suwannee River drains a large wetland, a kind of Venetian canal, or water garden, with bands of gorgeous mosses decorating the bottoms of the cypresses from which protrudes, for a fleeting moment, the head of a marsh hen, abruptly withdrawn.

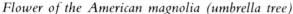

Flower of the American magnolia (umbrella tree)

Polyphemus moth lands on Virginia chain fern in Georgia.

The black, winding water looks planned by a landscape artist, especially at dusk, when the hollow grunts of cranes make the water seem to palpitate with dark and primeval mystery.

The yellow flowers of water lilies, the floating hearts, bloom in late October among the silent watching shapes of wood storks in one burned, dead cypress. Then, shock! A white ibis stands in a dark pine, as if sculpted from marble in green stone. Dozens of wood ducks are bursting upward from the water weeds. Bedlam, then peace, as they go down again.

The river winds among sentinels of herons, eyes dimmed with frogs, near mounds of cypresses in the middle of the river, sculpted gracefully up from its surface, ancient trunks now colonized by marsh marigolds, shrubs, ferns, and even bay trees. They look like miniature cities of plants, and, in the rushing light of dawn, urgent broad-winged feeders overhead, like fairy castles.

A silence of vultures rises on heavy wings, from watching places not seen by others, and is gone back up the river, into one of the hearts of the Southern wetlands.

In the fall, the birds are more muted, but they are displayed against backgrounds that are quite suddenly more vivid. The colors have changed, shuttling through swamp and prairie like New England leaf changes, but here in the South, in both richer and more subtle shadings.

Entering a Southern prairie, the traveller is flanked by banks of yellow and mauve flowers, orchids chancing a late meeting with early frost, against a scream of blue jays. The blue herons are everywhere, as are one hundred ibis scarring dead cypress with white wounds.

Old-man's beard is aloft, among the watching birds, but although it resembles Spanish moss, it is not a flowering plant but a lichen rooted in bark and getting its food from crevices, not from the bark itself; so it is not a parasite, merely a hanger-on.

The birds may be seen now in a different configuration of the morning rush to feed, thousands of discreet small figures moving towards where the gallberry, a kind of evergreen holly, is holding its black fruit for them.

Colors blaze afresh in the autumn, when so much of the wetlands have a decadent look to them, lustrous, but overripe with the humors of long summer. The sweet gum then takes on all the northern colors of fall, one of the largest and most beautiful of the wetland trees. Its colors march through yellow, ocher, burnt sienna, wine, crimson, purplish black, lavender, indigo, and cobalt.

114

Louisiana iris blooms in many shades.

A Northern cardinal stands on mistletoe.

Male summer tanager in bald cypress

OVERLEAF: *Virgin bald cypress stands among knees in South Carolina.*

120 *Carnivorous crimson pitcher plant attracts its diet of insects in the rain.*

Red-purple flowers of parrot-head pitcher plants in Florida

Yellow lady's slipper in the rain

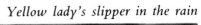

121

White-tailed fawn is hidden by the mother.

The white-winged birds sail into the clouds in flights that must be for joy, and beneath them, the peregrine strikes a straight line across the flowers of pipeworts in April, green orchids in August, swamp fetterbush in April, staggerbush in May, floating hearts in July, leather flowers in June, golden canna in July, swamphaw in May, deer's tongue in August, and Barbara's buttons in June.

The sweet gums are drilled with holes by the yellow-bellied sapsucker, and yellow-rumped warblers are eating the small, gray rounded fruit of wax myrtle if they have decided to winter in the Southern wetlands. The dwarf pawpaw's smooth berries are snapped up by raccoons and possums, its flowers cross-pollinated by beetles eaten by herons, its leaves scissored by caterpillars of the pawpaw sphinx moth, which are eaten by the secretive rails, and the adult moths hunted down by omnipresent bats.

123

Red-shouldered hawk perches in a tree.

Young killdeer walks among oyster shells in Louisiana.

There is mysterious presence everywhere.

There is mysterious presence everywhere, concealed in the secrecies of all biologies. The Spanish moss is mysterious, producing infinitesimally light seeds which are carried invisibly to new places of germination in the wind. They do not grow roots where they settle and, in fact, do not ever become a part of the tree which is hosting them. They are completely self-contained.

But the Spanish moss is unlike almost all other air plants, many of which store water or nutrients, attach themselves to their hosts, and produce fruit. The Spanish moss is the most successful of all the air plants, but it is not a moss, it has no roots, it bears no fruit.

Its food comes to it in rain, washed from the higher branches of its host plant; dead cells, flaking bark, all rich in minerals, are caught and held in tiny fine hairs on the Spanish moss stems.

The thrust of plant life, made movingly dramatic if any place of competition is visited days apart, is also the essential wealth of the Southern wetlands. A shrub seedling is crushed under the weight of a fallen cypress limb, but survives by winding its trunk, over scores of years, around the cypress again and again, until its sprouting tip finds freedom, and heads upward. Behind is a perfect replication of a coiled snake.

Thousands of cypresses are bent, contorted into shapes of agony, some of them horizontally at right angles, then bent upwards again at 90 degrees, bent and rebent, like stepladders. Egrets, or anhingas, perching clumsily on the leader tips of cypresses, break them off. The tips grow sideways until they can make their thrust upward again, only to be rebroken, the process repeated.

The abundance of plants is also the massiveness of death and decay. Every form of life thrives in this constant re-creation of raw materials back into gas, organic matter, protein, carbohydrate, phosphate, and acid.

When decay becomes monstrous, masses of bottom material rise to the surface in "blowups" which may be big enough to create new, floating islands, on which trees soon appear, and grow a score of feet. A hurricane comes, the island moves, and all trees fall down, to start the process over.

The American avocets leave for South America. The Canada geese leave for Saskatchewan. The snow geese arrive to spend the winter. The alligator catches a green-winged teal. The wood duck makes vivid the already bright leaves of the Louisiana fall. Solitary mauve flowers of the pipewort take in the colors of a sun falling into the Mexican Gulf.

Fading Days

In the moments before bright day becomes twilight, the Southern wetlands become a diversity of action and memory. It has been impossible to see, feel, record, know all but a small fraction of the total blaze of birds and plants from Florida to Texas. Instead, this later moment of the day is a time for synthetic summary, a rush of action before night can begin.

It may be remembered in many ways. The vignettes are as sharp as the winter sun hiding in magnolias. It may be a time for eggs. The eagle's pass over the wading birds' rookeries often sends showers of eggs, or nestlings, or both, plummeting down into alligator waters. It might be an alligator terrapin, its head almost as big as a man's, hauling its one-hundred-pound body ashore to lay its eggs.

It may be the Florida terrapin laying her eggs in an alligator's nest, her eggs timed to hatch earlier than the bigger reptile's. Eggs haunt the seasons, millions, billions of them. There have been eggs in hollow trees, eggs in delicate, basketlike nests, eggs hidden in hollow stumps.

Great egret stands over its nestlings.

The banks of the Suwannee Canal may seem plowed by machinery in the late afternoon of this early June day, hundreds of turtles finished their work of burying eggs, the hunting of raccoons and snakes to dig them up again not yet begun.

Or, in a sharp switch of atmosphere, the stage is occupied by young bears, born in January or February, who often get their start to adulthood by quietly spending their late afternoons and early evenings digging up such turtle eggs.

Late afternoon is the earliest moment of a new day—or night—for many. The bobcat comes to the edge of the prairie and looks fixedly, in daylight, at a heron standing in the shallows. This is the most territorial of all the long-legged birds and is, perhaps, the only one that eats, sleeps, and breeds all within a confined area of a single wetland region. The bobcat knows this and marks the bird's position at that moment of the day when the bird's vulnerability will become crucial in the swift approaching night.

It is the late afternoon which gives a sense of reprise to all lives in the Southern wetlands, because, perhaps, the sun is so often a falling inferno behind all vision. But sun or no sun, the sense of oncoming night is the essential atmospheric. The cypress timber lumberman of the Eastern wetlands, UHUH Hebard, was an intent watcher of the late afternoons. He once wrote, "My most unforgettable sight of an egret was one winging its way lone on November 25, 1932, against a background of dark clouds under the arch of a complete rainbow on Chase Prairie. The late afternoon sun reflected the green of the 'houses' of the prairie and the white of the egret's body and wings against the dark sky."

Jackson McGillvray, another wetlands walker, recalls a similar incident in August, 1931, in the Great Dismal, where he was watching a great blue heron, standing tall and motionless, with a small flock of ducks dabbling mud in the late afternoon.

"Suddenly," wrote McGillvray, "that great deep call of the giant heron rang out across the prairie. It was the clarion ringing, and every creature within earshot understood what the big bird was saying. The ducks flew instantly. Songbirds fled among the trees. I marvelled at the intricate connection of these lives.

"Then, almost belatedly, perhaps, the big heron took wing. I am sure he was not the real target of the men in the swamp on that day, but it did not matter. Such a giant flying target was irresistible. Shots rang out. The heron, a ruined sailplane, slowly crumpled and folded up in midair."

When night is close, and the division of different atmospheres of the wetlands touch each other, it is easy to recollect all that is vivid,

Sunset through cypress trees in Georgia

such as drought. And it is in the late afternoons of summer droughts that the wetlands seem ready to explode with the heat they have stored.

In 1931–32, drought was a catastrophe in the Eastern wetlands. Millions of acres burned and, as a result, thousand of cranes fled into the Florida peninsular, causing overcrowding and squabbling.

At the beginning of the drought, the Okefenokee was particularly hard hit. The wood stork, which was once a late summer visitor to the Eastern wetlands of the South, remained during this winter of 1931–32 because the fish also found themselves concentrated in the remaining alligator holes. Gerald Fortman, a New York ornithologist, observed that winter in three visits to the South.

"It was both marvelous and grotesque," he wrote in a personal diary note. "The mortalities were terrifying. Yet there seemed to be millions of fish, and both ibis (wood stork) and alligator seemed to

A Northern pintail drake flaps his wings and swims with hen.

have come to some kind of agreement to keep out of each other's way while they harvested the feast."

Fortman said the storks dragged the larger fish they caught, or the fish crippled but not eaten, to the shores, and when they had gorged themselves, their leavings were commandeered by dense flights of vultures which patrolled the swamp in hundreds, and, toward the later part of the winter, thousands.

The late afternoon is the time of the wild turkey, at least for this wary, bulky bird to give fleeting glimpse of itself. "I doubt if any wild turkeys remained in the swamp," said Hebard, referring to the early thirties. "They were killed for food by residents and not by wildcats." In the early 1930s, there were still flocks of twenty to forty, which made any hunter's fingers tingle with the possibility of bringing down a bird that was not only large but extremely succulent as well.

"They were hard to shoot because of deceptive speed, and had

A green-winged teal swimming

A snowy egret fluffs feathers near its nest.

Wild turkey struts on a chenier or high place in Louisiana marsh.

OVERLEAF: *Cattle egrets in flight in Bird City in Avery Island, Louisiana*

Eastern gray squirrel eats possumhaw berries.

It is possible to stand, and
just listen, and the wetlands
will reveal themselves
through voices alone.

Eastern screech-owl perches among possumhaw berries.

Bullfrog stretches in shallow water.

The vistas of the wetlands are uniquely,
almost aggressively, Southern.

A marsh fire hides the sun and may burn for days.

tremendous acceleration of takeoff. The birds weighed up to twenty pounds, and gathered in winter flocks of forty, going into the tallest pines to roost, coming in around 4 P.M. They were said to breed in March, hatching young late in April or early May."

The widespread drought of 1932 also brought in an invasion of great horned owls which became both visible and audible. Probably, these great hunting birds were picking off straggling water birds which were reluctant to leave their familiar territories in the wetlands.

One Southern wetland walker, Frances Harper, wrote of this time, "I noticed that when the deer came to feed in the late afternoon, the great owls begin the most indescribable series of caws, screams, and demoniac laughter, which made me realize that I was in the depths of a real wilderness."

The wetlands of the afternoons await, and the birds arrive strategically, just before night begins, and show themselves before passing on. The Northern parulas are transients some years, residents in others. The magnolias arrive first, but disappear from the water world at any time. Kinglets, vireos, and titmice come in waves and leave in ripples.

The American robins rush into the wetlands in flocks of hundreds just before daylight ends at the full moon and scatter, gorging at once on black gum berries. At the same time, bluebirds have filtered down from drier country in the North and provide delightful splashes of color at the end of almost any day in the Eastern wetlands.

No artifact or art of man is visible anywhere now.

Great egrets and tricolored herons build nests close together.

Farewell by Moonlight

In the spring nights, all movement in the Southern wetlands is against an uproar of voices which seek to outdo each other in claim of the season. It may be the screaming of the oak toads which reduce almost all other sounds to negative value, unless it happens to be the overmastering bellow of an alligator nearby. Then, behind the oak toads may sound an almost constant droning sound, the voices of Southern toads, the bass instruments in the symphony orchestra of the night.

White wings slash the moonlit air, and birds are gone to a secret destination which no photographer can recall. The gopher frog gives an immense hollow cry, as if of sorrow at the departure of the birds, and the pinewoods tree frogs bark back an answer.

Then, when the moon goes down, the wetlands are sepulchral with notorious promises, the hollow thundering of a solitary alligator, no reason for that one cry, perhaps made out of season. If a great horned owl gets into a rookery of egrets, the screams may be heard for miles and it may take half of the following day before the rookery is reassembled and the birds have calmed down.

A white-tailed deer "in velvet" drinks from pond in Louisiana.

Night has been heralded for hours, through those lambent late afternoon hours. The pall of swamp fires, the blazing of falling suns— night cannot be entered without fanfare. Sometimes, the sense of change is almost unbearable. The sky flames crimson, gold, green, yellow, pallid blue, fiery orange, and the colors run among the marshes, the hammocks, the prairies, the cheniers, the salt marshes, the Delta lands, the Everglades, the keys, the swamps. The last of the wild turkeys, crimson bullets, slip into their roosts.

For most wetland watchers, these are the tenderest hours of the twenty-four. Enigma begins at dusk. Not even the swamp people roamed these lands after dark without blazing torches, packs of dogs at their heels.

Blinded by the dark, the human animal can only listen. The birds, he may discover with some shock, have made the nights their own. Then, as blindness is admitted, it is possible to concentrate on the images of day as the reality that can be handled.

The birds can often be seen as silhouettes against the rising moon, silent, perhaps, before the night uproar begins, birds that may be flying from northern Oklahoma to Central America, birds that may be moving from western Georgia to the Atlantic shores.

At twilight, just as the frogs are hesitatingly beginning their night cries, the wading birds all delay their departures to roost for late chances to hunt one of their favorite foods. The herons walk stealthily among cottonmouths, bound for the same hunt. Ibis linger with young alligators, both species eager for frog. Anhingas sink to a strike of neck and foot when a red-shouldered hawk hovers in the last of the light, also waiting for his chance to stoop into water like an osprey.

Sandpipers and plovers and other shorebirds are flickers of quick movement in Everglades gloom as they come to roost on tidal mud flats. The purple gallinules are eager for the night to start, when they can make doubly safe their stalking search for the uproar of frogs that is growing on the far side of the inlet.

William Bartram was on the lower St. Johns River, in Florida in 1774, at dusk, and he was overwhelmed by the noise of early evening, finding it impossible to sleep. "The continual noise and restlessness of the sea fowl . . . all promiscuously lodging together, and in such incredible numbers that the trees were entirely covered."

Audubon was at Sandy Key, in 1832, also at dusk, when the birds of the wetlands were coming to roost, or rest. "The flocks of birds that covered the shelly beaches and those hovering overhead so astonished us that we could for awhile scarcely believe our eyes."

In the lustrous light of the Mississippi dusk, when Gulf light seems to play with low-lying clouds and gloss still waters with extraordinary colors, the bare bright-blue skin around the eyes of male

144

Olivaceous cormorants at nest in Texas

Four young raccoons crawl over a stump.

146

Female American kestrel perches in an oak tree.

Little blue heron raises its feathers in excitement.

147

Northern mockingbird perches on snow-dusted pyracantha shrub in Arkansas.

anhingas shows their readiness to breed in the spring days to follow. The American white pelicans, clumsily crashing to roost in mangroves on the keys of southern Florida, look like cargo planes against the dying light, showing in silhouette that extraordinary hornlike growth on top of their beaks that indicates their readiness to breed.

Dusk is a time for piracy as well as late hunting, when frigate birds, having nested in the Caribbean, come north to the wetlands of Florida to make their hunt, often trying to intercept home-going pelicans or gulls. The pelicans fly late, and often with beak pouches filled with fish, and the ground-watcher, already in deep gloom, may be astonished to see fish falling from the sky, disgorged by the frightened pelican, and the frigate bird not deft enough to catch his food falling into night.

For experienced watchers, dusk—even night—is the time to be abroad in the wetlands, when exotic species of birds are less wary, or more confused, and make themselves visible. In changes of season,

Male Northern cardinal eats possumhaw berries.

particularly when cold air sweeps out of the Southwest, bringing desert sparrows into Arkansas or Louisiana, or when Caribbean storms exhaust themselves in the east, and mangrove cuckoos appear suddenly, the Southern wetlands can become places of double mysteries, filled with strangers which cannot be seen but which may be known only by voice.

The dusk descends into mangrove and cypress, the air above is still light, making possible an almost ghostly observation of bird life. Such is the sharp demarcation of day and night in the far South that home-going birds may be seen from pitch darkness in a swamp, illuminated as if by brilliant white torches.

The swallow-tailed kites, which in midsummer may assemble in flocks of hundreds in readiness for Southern migration when their own breeding is completed, often put on displays of late-evening aerobatics which make them look like motion pictures playing out dramas on a dark screen as big as the sky itself.

OVERLEAF: Fog on pond at midday on Avery Island in Louisiana

Male anhinga spreads his wings to dry after diving for fish.

152

Black-crowned night heron wades and fishes.

Two limpkins (an endangered species) feed on apple snails in Florida.

154

Mallards feed at sundown.

The shorebirds give a kind of final flourish to the night because they are so frequently abroad in darkness. Sometimes this is because of confusion in migration, sometimes because it is the safe time for them to eat.

Through night glasses, American avocets cut across the Louisiana marshes, dark blurs in the darkness. Semipalmated sandpipers throng on a Mississippi sand bar, dark bunches of constant movement, settling for sleep. Elegant, graceful willets, with their penetrating calls—ill-ill-illet—take off into moonlight, showing striking black-and-white plumage, and scaring up a thousand other shorebirds in the Alabama marsh by the loudness of their voices.

Greater yellowlegs, curlews, ruddy turnstones, golden plovers, black-necked stilts, red-necked phalaropes, even parasitic jaegers, all may be a part of a night parade of birds, mostly unseen except as jagged chips of sound carved from blackness.

If there is final farewell to the wetlands, it is given by creatures at opposite ends of the scheme of evolution. The spring departure of waterfowl, up the central flyway, to the middle states, Canada, and the low Arctic, is one of the spectacles of nature that none can forget.

These millions of striving bodies, suddenly aloft in moonlight, are a shock to all sensibility, because they are both thunder and lightning, the boom of their wings counterpoint to the moonlight writhing on their lashing wings.

Then, far behind them, in sepulchral acknowledgement of the distance in time that separates them, the reptiles also assert themselves. The bellow of alligator voices is the last sound of the Southern wetlands before distant dawn is suggested. When the sun appears again into a silent aquatic landscape, pulses quicken everywhere and a unique world pauses to waken.

The birds can often be seen as silhouettes against the rising moon, silent, perhaps, before the night uproar begins.

157

Great egret feeds her nestlings.

Great egret in flight

Wings on the Southwind

Designed by Bob Nance
Mechanical art by Design for Publishing
Homewood, Alabama

Text type is Linotron 202 Bembo by Akra Data, Inc.
Birmingham, Alabama

Color separations by Graphic Process, Inc.
Nashville, Tennessee

Printed and bound by Kingsport Press, Inc.
Kingsport, Tennessee

Text sheets are Lustro Offset Enamel Gloss by S.D. Warren Company
Boston, Massachusetts

Endleaves are Multicolor Corduroy by Process Materials Corporation
Carlstadt, New Jersey

Cover cloth is Arrestox Linen by Joanna Western Mills
Kingsport, Tennessee